HOW DOES
THE HUMAN BEING
MOVE?

THE PROBLEM OF
THE MOTOR NERVES

L.F.C. MEES, M.D.

Originally published in Dutch by the author in 1988
Translated from the Dutch by Philip Mees

Afterword by Douglas Roger, M.D.

Copyright © 2012 of this translation
by Mercury Press

ISBN 978-1-957569-04-8

Printed and Published in the USA

MERCURY PRESS
an imprint of SteinerBooks
PO Box 58
Hudson, NY 12534
www.steinerbooks.org

TABLE OF CONTENTS

Introduction 1
A Surprising Statement 5
The Riddle of Human Movement 8
Preliminary Review 11
Movement and Perception in Animals 14
Movement and Displacement 18
What Lives in our Muscles 21
How Movement Comes About 25
The Connection with Social Life 30
Consequences 35
Afterword 39
Addendum I 47
Addendum II 53
Addendum III 61
Notes 63
Bibliography 65

Introduction

In 1914 the First World War began. I believe we are justified in calling it an economic war. Because I experienced it quite consciously, I remember very well how one of my high school teachers, convinced that Germany had wanted this war, said to us: "Germany has already conquered the world." An example of this conquest was the well-known label 'Made in Germany' on their products, which even the English had to admit were superior to their own.

The peace accord of Versailles in 1919 was hard for Germany, so hard that it was soon evident that the stipulated reparation payments would drive Germany deeper and deeper into debt. Without any doubt this was the objective of the 'tiger' of France, Clemençeau. The Americans provided a solution, but we need not go into that here.

It will be interesting for many people to hear that Rudolf Steiner, convinced of the great injustice done to the German economy, said the following "...either people will accommodate their thinking to the requirements of reality, or they will have learned nothing from the calamity and will cause innumerable new ones to occur in the future."[1] In the background of this statement stood a law he perceived in social life, which he elaborated in his book *Towards Social Renewal*. Isn't it important to recognize that he precisely hit the nail on the head? Only twenty years later the Second World War began! Here Germany also took center stage. This war had primarily an ideological background. While in the first war the goal was to limit Germany economically, in the second one it was a defense against an ideology of which many of us have experienced the bitter consequences.

The two concepts 'economy' and 'ideology' reveal a profound contrast. When we think about the concept of ideology, we find that this is most prevalent in the eastern part of the world. In the West we find most of all economic problems. Thus we have Ideology in the East – Economy in the West. Let us take a closer look at this contrast. With ideology we run the risk of developing intolerance which, in turn, brings the dangers leading to dictatorship. We know all too well how this has worked in the past. Communism can be viewed in first instance as an ideology, and the concepts of intolerance and dictatorship are by no means unknown in Russia.* By contrast, the economy will always entail competition which, in turn, leads the human being to develop an increasingly egoistic view of life. So we see:

        Economy – competition – egoism
and
        Ideology – intolerance – dictatorship

Egoism and dictatorship emerge here as the greatest contrast! They characterize the concepts 'individualism' and 'totality' (as used in 'totalitarianism').

We can go one step further and say that dictatorship will always be indissolubly linked with a specific persuasion which can best be characterized as dogma. To the outside world, dictatorship will manifest first of all as power. The egoism of economic development will all

---

* The author wrote this in the mid 1980's (PM).

too easily lead to a materialistic view of life. One can also regard 'materialism' and 'dogma' as polarities. In its external manifestation materialism will foster a striving for wealth. Thus we arrive at a further contrast, namely between power and wealth, the two principal trump-cards of East and West:

> Economy – competition – egoism – materialism – wealth
> and
> Ideology – intolerance – dictatorship – dogma – power

Naturally there are also economic problems in the East just as there is also the will to power in the West, but the deeper essence lies, it seems to me, in the contrast indicated above: power in the East, and wealth in the West. We can also see that these two tendencies clash with each other – hard, as hard as stone, the stone of the Berlin wall. I cannot think of a more miserable, more serious calamity in the world than the existence of this physical barrier.

Earlier in this introduction I mentioned Rudolf Steiner's book *Towards Social Renewal*. We will return later to the threefold nature of the social organism which is the subject of that book. For now we need to look at the number 'three'. The East-West contrast can only be reduced to its proper proportions if a third principle, a Middle, a Center, can arise – a Middle in which the two polarities can meet. Only then the possibility would arise for the development of a true social life.

The error of original communism is, in my view, that one wanted to *prescribe* social life. That means that it can never really be born. But the increasing egoism in the West is just as much of a barrier for the development of social relationships. We only need to think about labor problems involving wages and strikes to see the effect of this.

What is the cause of this? Why is it so difficult for humanity to develop a sense of social awareness?

A Surprising Statement

Early in the twentieth century Rudolf Steiner made an extremely noteworthy statement regarding a particular part of human physiology. He said that all nerves are sense nerves, and that there are no motor nerves. This becomes even more interesting when we pay attention to the sharpness of the words he used. For he spoke of the nonsense of motor nerves, which in my opinion clearly indicates the self-evident fact that these motor nerves are in reality sense nerves![2] In physiology the word 'sense' indicates the capacity to perceive. Rudolf Steiner said of the so-called motor nerves that it is their purpose to perceive our movements[3].

Totally surprising is his conclusion that if humanity will not learn to recognize the nonsense of motor nerves it will not be able to develop a proper social life.[4]

Whoever reads this for the first time will probably be rather perplexed. I suppose there is not a single physician or physiologist in the world who has ever doubted the existence of motor nerves. At least, I am not aware of any. It is only a certain confidence one learns to develop in the course of time in the person of Rudolf Steiner that can save us from rolling our eyes at such a statement.

It is perhaps good to insert a few words about the life and work of Rudolf Steiner here, although it will have to be very brief. The only thing I allow myself to write down is that Rudolf Steiner was so deeply familiar with all areas of science that it astonished many people in his day. True, these were only those who came to know his works. (And it may be said that his work belongs to the most voluminous that anyone ever left behind.) The fact that many people do not make the effort to become acquainted with it is, unfortunately, based primarily on prejudices both religious and scientific.

When someone does take the trouble to read something by Rudolf Steiner I would advise him or her to ask not only *what* he wrote but also *how* it is expressed. Since I knew Rudolf Steiner for several years I am speaking from experience when I say that I cannot remember ever having encountered a more modest person than he was.

One can characterize anthroposophy[5] as the development of a new view of our environment, which not only engages our head but also our heart. Despite much resistance, anthroposophy has shown its fruitfulness in many areas, including pedagogy, agriculture, medicine and the arts. This is demonstrated by the steady growth in the number of Waldorf Schools, biodynamic farms, anthroposophically-oriented physicians, hospitals and therapeutic organizations.

I have made these remarks only to indicate that Rudolf Steiner's enigmatic statement mentioned above can at least be taken seriously enough to wonder what the pronouncement that there are no motor nerves may be based on. And how can it be that official physiology and medicine have been so completely blind to this, and still are?

Rudolf Steiner made this statement not once but many times. In the bibliography the sources are listed where these statements were recorded. In a lecture on June 8, 1919 in Stuttgart he said the following:

A sensory nerve, a sense nerve, exists as an instrument for us to perceive what takes place in our sense organization. And a so-called motor nerve is not a motor nerve; rather, it is a sensitive nerve; it is there only so that I can perceive my own hand movements, my own movements, which come from sources other than motor nerves. Motor nerves are inner sense nerves for perceiving the initiatives of my own will. In order for me to perceive the outer world, to perceive what occurs in my sense apparatus, sensory nerves exist. And so that I do not remain an unknown entity to myself when I walk, strike, or grasp without knowing anything about it, the so-called motor nerves exist. So motor nerves are not there to activate our will, but rather so that we can perceive what the will does in us. [6]

And his comments regarding the connection between this insight and the development of social life go on to indicate that if this insight, namely that the motor nerves are sense nerves, does not find acceptance, humanity will not be capable of developing an appropriate relationship to its social environment.[7]

The Riddle of Human Movement

The Conventional Theory

Perhaps many people will be surprised by the fact that I am speaking of a riddle here. They may say that it is self-evident that our muscles are the ones that move our body. Muscles are able to contract. They are connected with our skeleton, which is almost completely surrounded by muscles. Because of this we can change the position of the various parts of our skeleton relative to each other by contracting our muscles – in other words, we can move.

The next question would be how this contraction comes about. About three hundred years ago it was discovered that there are supposed to be two kinds of nerves: the sense nerves and the motor nerves. Sense nerves connect our sense organs with our brain. From our environment we receive a variety of impressions, such as light, sound, etc. These impressions are sent to our brain through nerves; they go from the periphery to the center. In science these are called 'afferent' nerves.

The so-called motor nerves have the function of transmitting impulses from the brain (or the spinal cord) to the muscle; they bring about the contraction of the muscle.* Since for making complex movements an incredible number of larger or smaller muscles have to receive stronger or weaker impulses to contract, the

---

* A strong argument for this theory is the fact that when this nerve is cut the muscle can no longer contract; movement is no longer possible: we are witnessing paralysis.

number and nature of these impulses has to be unimaginably large and varied. When a human being wants to execute a certain movement he will, according to this theory, send impulses from his brain to the muscles through the nerves. These impulses travel from the center to the periphery – they are 'efferent' nerves.

When the perception through the sense nerves is described it is also necessary to answer the question of how the impressions that are brought to the brain through the nerves are transformed into the content of our soul life. This is a question which science to this day has not been able to answer. And when we ask where the connection lies between our soul life and the impulses that travel from our brain to our muscles when we move, science again is silent.

In an older representation of the human being, showing a picture of the human being as a 'Palace of Industry', we see in the brain a number of operators directing the processing of machines in the body in which one could think that the various movement (motor) impulses were being sent to the muscles.[8] In some way or other these operators take the place of the human soul in this picture. But who these operators are and where their instructions come from to operate the machines properly is not explained.

Science has always tried to find examples to test such hypotheses through experiments. A well-known one is the experiment with a frog muscle one end of which is suspended from a fixed point, while the other end is tied to a little lever. In medical terminology this set-up is called a muscle-nerve preparation. When one now stimulates, or irritates, the nerve of this muscle in some way, through electricity, squeezing, heat or whatever, the muscle gives a jerk and the lever is pulled up. Of course, the muscle then becomes shorter, thicker and harder.

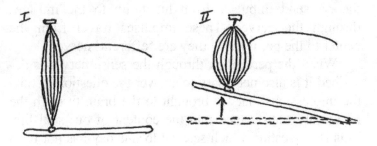

In this way one tried to show in a simple manner how movements come about. The lifting of the lever was called an 'elementary movement'. In the study of medicine we get to know the same phenomenon by causing so-called reflex movements. Everyone knows the reflex of the knee tendon. When someone crosses his legs so that one leg hangs loosely over the other and one then taps the knee tendon, the leg will 'kick out' a little. Many people know this from their own experience.

Here one could also speak of an 'elementary movement'. And movements in ordinary life, although vastly more complex, are thought to be based on the same principle. We need to add to this that, according to this approach the movements of the human being are directed from the brain. The moving human being could thus be viewed as a puppet of the brain or, if you will, of the head. When I asked several people what they thought of this conclusion they fully agreed with it.

Preliminary Review

At this point it becomes necessary to take a look at the way some people think Rudolf Steiner's statements should be interpreted. We have seen that, according to 'official' theory, it is assumed that through the motor nerves a stimulus is sent to the muscle. Contrary to this, however, it was thought that we should try to imagine that through this nerve an impression is transmitted from the muscle to the brain, the way it goes with the sense nerves. It was, therefore, pictured that the 'flow' in the motor nerve had to be imagined *in the opposite direction*.

Rudolf Steiner's second point will then be understood in the first instance as: the human being 'recognizes his movements' because he becomes conscious of what is taking place in his muscles. The reason that after the 'motor' nerve is cut the muscle is paralyzed would then be the fact that we can no longer perceive our movements.

But is this correct? Are we conscious of our movements because we notice what is taking place in our *muscles*? Perhaps someone will mention as an example how we feel the tension in our muscles when we lift something heavy. But then we must not forget that the tension in our muscles occurs because of our *inability to move* properly! When I want to push down a tree there is enormous tension in my muscles, but absolutely no movement takes place. And on the contrary, with a normal, unimpeded movement we are *not* conscious of what the muscles are doing. How then do we really know that we move?

Of course, we can *see* that we move, but in order to discover the precise 'seat' of the perception of movement one can experiment by making tiny, minimal movements with one's fingers. I have often asked the audience at my lectures to do this and then asked them where they became conscious of those movements. In most cases it took a little while before someone suddenly came with the answer: in our joints. Especially with very small movements of the fingers one can feel quite precisely that we become aware of them through the joints of the phalanges. Of course, that which takes place in the capsules, the tendons, the skin, etc. also contributes to our awareness of our movements, but we are justified in saying that our joints are the principal 'seat' of what we commonly call our sense of movement.

And now, what about the idea that we would be paralyzed if we would no longer be conscious of our movements? There are diseases of the spinal cord because of which one does not receive impressions from the joints. People who suffer from these diseases do not know that they move, but they do have the ability to move. Because they do not perceive their movements in the normal way their movements are too large, lurching, and they constantly have to try and control them through their sense of sight, touch and other senses. This is called ataxia. This means that, even though we have no consciousness of the movements themselves, we still have the ability to move; we are not truly paralyzed!

The fact that, as mentioned above, we are completely paralyzed if the so-called motor nerve (which I will hereafter call muscle nerve) is cut, is in my opinion not satisfactorily explained by the unconditional statement: we can then no longer perceive our movements.

It has to be emphasized, though, that we have been dealing here with a *particular interpretation* of a statement by Rudolf Steiner. The point of departure for this interpretation was that motor nerves can only be considered as having a sense function if they would be *afferent nerves in relation to the brain*, i.e. that the flow in the nerve would go from the muscle to the brain.

The meaning of Rudolf Steiner's statements about the motor nerves has to be that we would have to break with the conventional representation that our muscles are made to contract by so-called motor impulses. The statement that "there are no motor nerves" then becomes the point of departure for a new consideration of human movement. Thus it now becomes our challenge to answer the question of what does cause the muscles to contract. After all, in one way or another there has to be a connection between me and my muscles.

Different people have come up with different ideas on this. Some say, for instance, that the connection of the soul, of the 'I' (after all, this must be the origin of the movement impulse) does not lie in the nerves, but in warmth. But even if we think that we bring about the contraction of our muscles through a medium other than our nerves, what have we really gained? We have said that in the old picture the human being is, as it were, a puppet of the central nervous system. In this new picture he is just as much a puppet, only in lieu of 'nervous system' we have to substitute the word 'I' or 'soul'. I wonder if this is helpful, because it still does not at all answer the question of how the contraction in the muscle is brought about. And finally, where is the connection between this idea and the problem of creating a social community?

Movement and Perception in Animals

We will now first concentrate on the question of what we can learn from the fact that by stimulating the nerve the muscle contracts. Let us therefore first ask ourselves: if I am in a half-dark room and I see something lying on the floor, how can I know whether it is a live animal? By touching it. If it begins to move I conclude that it is an animal. Why does the animal move? Because it has perceived something. How do I know that it has perceived something? Because it moves. Why does it move? Etc., etc. Perception and movement are the only things I can assess. What takes place between this perception and movement completely eludes my observation. Or perhaps not completely?

When I step on the tail of a dog, why does the dog yelp? Virtually everyone will immediately answer: because it has pain. How do we know that? After some hesitation we usually get the answer: because I would do the same if I were in pain. That means that we have at least some notion of what an animal feels. The main point is, however, that from this single phenomenon we can derive a rule, namely that movements of animals are always accompanied by some form of perception.

Let us now imagine that while digging in our garden we accidentally cut through an earthworm. The animal will wiggle and squirm, also the hind part. Why does it wiggle? If we say that it has pain, I would have to point out that the little word 'it' is not valid for the hind part, which is no more than a piece of the worm and, importantly, lacks a head. Nevertheless, everyone realizes that it would not be right to say that no perception was taking place. The impression the squirming movements make on us speak eloquently: there is perception. The fact that movements of animals are accompanied by perception is always valid. Of course, I cannot deny that the pain may not be the same in the worm as in the dog. But I say that it is totally justified to state that in the piece of worm *perceptions* are taking place.

When I was a young boy in Holland there was a fishmonger who came through the street offering fish door to door. He had on his cart live eels in sand. As children we would often gather around him when he would cut an eel close to the head and then with a quick, practiced movement strip off the entire skin and cut the eel into pieces. You could clearly see every individual piece twitching.

Why these little motions, this twitching? Because perception was taking place! From the fact that movements in animals are always accompanied by perception we can now learn that an animal is a perceiving thing, and also a piece of animal can still have perception, at any rate for some time!

I can understand that, since there is in modern science no adequate understanding for this principle of perception, the principle of becoming aware, this will present a difficulty for many people. But I believe that this is unavoidable. Yes, we would even need to go a step further, and describe for the human being a third form of life: human self-conscious life. In my opinion, the French scientist/monk Pierre Teilhard de Chardin characterized the difference between man and animal in a brilliant manner as follows: "The human being knows and the animal knows; however, the human being is the only being that knows that it knows."[9] The concepts of 'plant life', 'animal life', and 'human life', as they are intended here, in fact stand for the words 'life', 'soul' and 'spirit' from daily life. Perhaps one can sense that we are pointing here also to the difference between what is called strictly scientific on the one hand, and on the other hand a range of ideas in which we also speak of realities we call life, soul and spirit, and which by their very nature fall outside the realm of exact natural science.

The series of dog–earthworm–eel can be extended by adding the example of the muscle that contracted when the nerve was stimulated. Why does the muscle contract?[*] We have to do here with the exact same phenomenon as with the pieces of eel and the cut earthworm: perception is occurring. As soon as we see this we recognize that the contraction of the muscle is a reaction to perception, awareness. However, *if the muscle perceives, then the muscle nerve is a sense nerve, a sense nerve **of the muscle**, not of '**us**'*. The nerves are therefore indeed *afferent*, but relative to the muscle! In other words, not 'I' perceive, but my muscle perceives! Thus it is not a case of having to reverse the 'flow' through the nerve, as some people thought.

We are clearly facing a totally new situation. We will see where it leads us. On the one hand I am a being who has perceptions through his sense nerves. On the other hand I have discovered that there exists a 'muscle being' in me that also has its nerves and perceptions.

---

[*] One almost wants to say: when I unexpectedly touch someone, he may be startled. In the case of the stimulated nerve we could say: the muscle is startled.

But for now I have no access to these with the conscious side of my being!

The duality of self-conscious 'being' and muscle 'being' points to two worlds in us. I am myself a perceiving being and I possess, in some way or other, a muscle system with its own life of perception. It is obvious that our first and principal task will be to come to an understanding of the way in which our conscious movements come about.

Movement and Displacement[*]

Now we need to revert for a moment to the experiment of the frog muscle; by stimulating a nerve a muscle is made to contract and, as a result, the little lever that was tied to the muscle is lifted. This was described as an 'elementary movement', but this conclusion rests on an erroneous interpretation of an observation. For it is not correct to say that the lever is *moved* upward by the muscle – it is *shocked* upward. It is decidedly misleading to speak of a movement here. No movement is occurring in the true sense of the word; rather, the lever is *displaced*.

Similarly in the case of the knee reflex that was mentioned as another example, we called it also an 'elementary movement'. But here also it is obvious that this too is a case of displacement. The funny, powerless feeling we have when we do this indicates already that we cannot speak of a movement here.

We have already mentioned that in modern physiology the human being is viewed as a puppet, in which our movements are deemed to be an incredibly complex totality of the kind of elementary movements mentioned above. I now have to correct this sentence and say that these are not movements but displacements. In this system of thought the moving human being is therefore not a moving being, but a being that is in fact constantly

---

[*] The word 'displace' is here used strictly in the sense of 'put out of its place' or 'change its place'. No additional connotation is meant. (PM).

displaced – in truth, a puppet! Naturally, we must now ask the question of whether moving and being moved are the same thing? Absolutely not! Something *is displaced*; something *moves* on its own! Of course, a movement also involves a displacement. But that does not mean that a displacement is a movement.

To illustrate the difference, let us look at the following experiment. When we draw a circle on a board, put a piece of gravel on the line and then try to move the gravel precisely along the line by tapping it with a little hammer, is that possible? Of course not. With each tap of the hammer the piece of gravel can only be displaced in a straight line. No matter how lightly we tap, the gravel will never precisely follow the line of the circle. The displacements always happen in straight lines. Perhaps one can sense that we encounter here the same thing as when it is said that human movements are built up from an infinite number of displacements, caused by an equal number of muscle stimulations.

How would we be able to cause the piece of gravel to move exactly along the line of the circle? By taking it in our fingers! As soon as we say this we find ourselves in an entirely different world. We no longer have to do with hammer and displacement but with hand and movement. By making these two experiments we can experience bodily the difference between movement and displacement: movement is a *gesture*, a gesture through which an action is born.

If I replace the piece of gravel with a heavier piece, will I still be able to move it around the circle? Of course. In itself, weight has nothing to do with the movement, provided I have enough strength. The heavier the piece is, the more strength I will have to develop, and the more I will have to use my muscles. Here we come upon a characteristic of the muscles to which we have so far paid little attention. More than anything else, muscles have to do with strength, force. It has also been said already that when more resistance has to be overcome in a movement, more muscle strength is needed. We must therefore make a distinction between two things: on the one hand the movement itself and, on the other hand, the strength needed to execute the movement. The movement is possible if the object is lifted out of gravity by the strength of my muscles.

When I do not move a piece of gravel but just make a movement with my hand through the air, there is still something I have to lift out of gravity: my arm itself. My arm is a heavy thing, but when I move it in daily life I notice nothing or next to nothing of this weight. And with this sentence something is stated which takes us back to the expression: 'I' move my arm.

'I' move my arm, not my *muscles*! We will now direct our attention to this point. What role the muscles play here is something we have already discovered in the experiment of the piece of gravel if it is heavy. Muscles have the ability to lift something out of gravity because they can develop *strength*, as we have said above. In what follows we will elaborate on this, but first I would like to say something that can deepen our sense of this whole picture.

I have used the word 'strength'. What is strength? Where does it come from? It comes somewhere from some hidden domain, and can fill us with a feeling of respect. Whenever the word 'strength' is used we often undergo a feeling of respect. For this reason we tend to respect an action, a deed, more to the extent that the person has mustered more strength to accomplish it!

What Lives in our Muscles

In my youth the way people danced was different from the way young people do now. We danced with each other. We held each other while dancing a foxtrot, a tango and especially a waltz. In those days the male partner was always the leader, the female followed. Those girls were the chosen ones who danced so lightly that you could hardly feel them. We said of them: "They follow so well."

How did they follow so well? Because they reacted to the slightest change of direction of their partner. One could say: they 'listened' to the intention of the other. One did not feel them, as opposed to those who were not able to 'listen' that well and were as heavy as lead.

We now need to make one more little step in order to see the riddle of human movement in an entirely new light. Just as we danced with our partner through the ballroom, in the same way we 'dance' with our skeleton when we move by ourselves! In this picture we have to imagine our skeleton as clothed in our muscles. Why do I not feel my skeleton? Because it 'follows' me just as my dance partner was able to follow me.

When I say that 'I' move my body I create a problem. For who is 'I'? A spiritual being! How can this spiritual being handle a heavy body? Only because this body is lifted completely out of gravity. My muscles do that. With every movement my muscles 'offer' me my skeleton (my body). We dance through life with our skeleton.

My muscles can only offer me my skeleton because they continually 'listen' to me. Therefore, in one way or other they have to have a connection with me that gives them the ability to do this 'listening'. This connection is formed by the 'muscle nerves' – formerly called 'motor nerves' – through which the muscle being is connected with me through the spinal cord and the brain.

The relationship of this so-called central nervous system to our soul life is, as we have already noticed, a problem in itself, to which we will return later. For now it suffices to state that nerves *always* serve to perceive, whether I perceive or my muscles perceive. Thus we come to the revealing thought that we possess in our body a second entity, which we have called our 'muscle being', and which leads its own autonomous life in me, listens to my movements and at every moment during my movements offers me my body. This being serves me. Yes, when I am tired, i.e. when my 'muscle being' is tired and is unable to develop enough strength, we appropriately say: 'my muscles refuse to cooperate'. Only then do I begin to feel my muscles!

The picture that my muscles move my body is closely related to this. With certain movements we can feel very clearly that we need our muscles to make those movements. We could even say that when we have to exert ourselves in a movement we begin to notice how we are straining our muscles. This means at the same time that we then also feel restricted in our movement.

When we lift a very heavy object, and thus make an extremely slow movement, we could say that this is an illustration of the fact that my muscles do this or, better formulated, that we do this with our muscles. But we must not forget that when lifting something in this way we can hardly speak of any movement at all. And to the extent that there is some movement, this does not contradict the statement that it is not the muscles that move my body, but 'I' move my body.

Always when we have to do with strong people, such as body builders or weight lifters, it is the strength that counts, not the movement. This is why I really like the picture of movement as *gesture*. If I would be asked whether in weight lifting a movement takes place I would answer: yes, but hardly. After all, strength and movement are opposites. The movement is made by **me**, I cause it, I form it in my gesture, and my muscles make it possible not by obeying me but by serving me. The fact that the muscle is paralyzed when the muscle nerve is cut becomes perfectly understandable in this picture. For the cut prevents the muscle's ability to 'listen to me'. It can therefore no longer 'serve' me. To live into this picture we must not start from exceptional situations as described above.

Perhaps we find the free, unimpeded movement best of all in our gesticulating. Here form and service meet in a free gesture in the most beautiful way.

Now we can add to all of this another surprising statement. My muscles listen not only to the movement impulses that proceed from my conscious 'I', but also to those that out of a completely different world want to steer my actions in a particular direction. For a person who is deeply convinced of the reality of karma, in other words, of the fact that we are guided then and there in connection with encounters that are only possible then and there, it will be a question of how this happens.

What is it that ensures that I go to that particular place where I can meet someone, or go to that particular place where I have an accident? Why do we often say, when someone has met someone we also know: isn't it a small world? No, the world is not small, but the group of people who have something to do with each other and who meet each other is relatively small.

These expressions clearly show that most of us do often have some notion of the reality of reincarnation and karmic connections, which lead us somewhere. However, the meaning of such events may reveal itself only later, although in most cases we do not become consciously aware of this. Something guides us to some place, but that means that there has to be a mysterious connection between that which guides us and my 'muscle being', just as is the case with conscious actions. My muscle being 'listens' to the movements which I want to make, but it also listens to the guiding impulse of my destiny, my karma.

The above statement that my muscles also 'listen' to the impulses from the spiritual world, which are connected with the mystery of our karma, could be an answer to this. Many people will feel that this is a more or less enigmatic way of saying things, but we can experience here how difficult it must have been for someone who lives completely in the reality of the spiritual world, to find adequate expressions in the material world to express what he means.

How Movement Comes About

In the "Preliminary Review" section of this book a connection was mentioned between the 'I' and the muscles through warmth. What was meant was that one tried to look for the cause of the activity of the muscles in the self-conscious 'I' which, because it lives in the warmth organization, was thought to have a connection with the muscles. Warmth, as connector of the 'I' with the muscles, was then used in an effort by some people to explain the contraction of the muscles. At this point, however, we want to follow the 'I', where movement originates, to its point of contact in the body. The task will be to answer the question how, according to the train of thought developed here, a movement comes about. For we have to come to an understanding of the way the muscles arrive at 'perception'.

The connection of the 'I' with the body will certainly have to begin in warmth. From here, one can imagine how, along a path of condensation (the air-like and the fluid-like), the 'I' must also be able to intervene in the deeper regions of the body. But here we come upon a difficulty. I am a spiritual being, my 'I' is a spiritual reality that seeks, through warmth, access to the 'less spiritual'. This has to have a limit. We can imagine that when I try to move my arm I encounter that limit in the action of taking-possession-of-my-body, namely there where I have to enter into interaction with gravity.

The muscles perceive this moment, and this is the moment when the mystery is enacted which is the subject of this book.

In *Fundamentals of Therapy*, written jointly by Rudolf Steiner and Ita Wegman, we come across a remarkable sentence. It goes as follows: ... *every movement is a beginning paralysis, which is lifted again at the same time.*[10] A movement begins as a spiritual activity which, because it has to take my body along with it, enters into interaction with the physical. The subtle instant when this is perceived by the muscles, which can then offer the intended assistance, can make the above quotation comprehensible.

We have been discussing a realm in our body that is full of activity of which, however, we perceive nothing. It is that which lives in our muscles and creates the possibility that we can make movements. What is it that is active here? To which element in our soul life might it relate? Where is something taking place of which we consciously know nothing?

Rudolf Steiner gives us a clue here with a most unexpected revelation. He stated in many different contexts that in his thinking the human being is awake, in his feeling he is dreaming, and in his will he is asleep. At first sight this is an incomprehensible statement. Don't we countless times express our will each day? We are certainly not asleep then, on the contrary. The secret is, however, that this kind of will is not real will. No matter how surprising this may sound, when we express our will we actually utter a *desire*, which is really still part of our thinking and feeling life.*

What then is it to *will* something? In my opinion there can be only one answer. Truly willing is *doing*!

---

* In English this is more evident than in German: we say *I want*, not *I will* (PM).

Perhaps someone will say: 'but I know what I do, don't I?' Most certainly, but we don't know *how* we do it. What is going on in our muscles, nerves, etc. as we perform an action escapes us totally. We have even pointed out already that the only way we experience something like awareness of our muscles is when we meet with strong resistance. In that case, however, movement is at the same time impeded.

Not knowing what is taking place in our muscles is for me identical with the statement that the human being is asleep in his will. This must mean that the human being is not conscious of his will. We have to do here with a different 'I' than when we say 'I am'. Thus we could say: in his thinking the human being has an 'awake I', in his feeling a 'dreaming I', and in his will a 'sleeping I'. This 'sleeping I' is, however, asleep only in relation to the 'self-conscious I'. In relation to something else it is not at all asleep, namely in relation to my movements. In my movements this 'I' is extremely active. For it continually listens to my movements, it follows me and continually offers me my skeleton (my body).

Here we have again reached the point that was the purpose of this entire study, namely to demonstrate that the muscle nerves are sense nerves. An obvious question will be how my 'muscles' become conscious of my desire to move. I believe we would do well to view what we call our nervous system as no more than the physical expression of what lives everywhere in our body as perceiving activity.

To the extent that we are conscious – or are able to be conscious – of impressions from our body, we have to do with that part of our nervous system that is always considered to be part of the realm of sensing. Similarly,

27

that which used to be labeled 'motor nerves' will now have to be called 'sense nerves'; they enable our 'muscle being' to become conscious of our movements. Since this muscle consciousness is, after all, also a part of me – which gives us the right to speak of an 'I' also in this context – it follows that Rudolf Steiner's sentence: *motor nerves are inner sense nerves for the perception of my own will's decisions*[11] is entirely correct.

The question may be asked why Rudolf Steiner did not say this more clearly. This study of the riddle of human movement can give a definite answer to this. Rudolf Steiner gave us riddles like this many times. For him the point was not just to communicate bald facts that we would be able to take into our life of imagination as such, besides all the other experiences in our daily lives. His intention was a schooling of consciousness. What we experience in our thinking when we, so to say, turn a particular train of thought completely on its head, illustrates what is meant here.

Just the expression '*I* move my body', together with the picture of my muscles helping me with this movement by offering me, a being performing an action, my body at every conceivable instant, requires a totally new insight. The modern view of movement with muscles, motor nerves, etc. is completely materialistic. One thinks one can make movement comprehensible by using the experiment with the frog muscle as a starting point.

When discussing this approach in the beginning of the book we mentioned that science has (as yet) no answer to the question of the relationship, which has to exist between an imagination of an action, with the impulses that go from the brain to the muscles. At that time we also said that this is equally true for the question of how impressions end up in the soul and spirit through the sense organs, the nerves and the brain.

But does the modern scientist suffer from sleepless nights because this problem has not been solved? Not at all. He may say that we don't know that yet, but one thing is certain: the impressions that come in from the outside (sense nerves), or that go out from the inside (motor nerves), must be able to be expressed in terms of molecular, genetic or biochemical processes. Anything is acceptable as long as no reality of existence is attributed to soul and spirit.

The concept of 'I move my body' makes a complete break with this approach. It raises the question in me: how can I, a spiritual being, the reality of which I unconditionally accept, move a material body? Taking the reality of the spiritual nature of the human being as a basis, this study opens a door to a solution and, most of all, to the value of Rudolf Steiner's few remarks on the subject.

We have now arrived at the point where we need to occupy ourselves with the third riddle Rudolf Steiner presented to us namely that if the insight that motor nerves are sense nerves does not find acceptance, humanity will not be capable of developing an appropriate relationship to its social environment.

The Connection with Social Life

We have seen that in the two realms, 'we' and 'our muscles', with which we are now becoming acquainted, one 'serves' the other. Here we come upon terminology we also use when speaking about social life. In a community, however, we are not looking for a situation in which there is only one who takes initiatives while another serves him or her, but that all people need to have both qualities at the ready all the time, if they want to work together. One could say that this reveals itself even in conversation, in speaking and listening, in other words, in the leading and following that constantly alternate.

The extent to which this study relates to problems in social life, such as this has developed in the course of the past few centuries on earth, will become clear in the following.

Many readers will be acquainted with the fact that in the anthroposophical view of the human being the threefoldness of this being, in both body and soul, plays an important role. Starting around 1916 Rudolf Steiner described the composition of the human body as consisting of a nerve-sense system, a rhythmic system and a metabolic-limb system. Similarly, the soul life of the human being is described in terms of thinking, feeling and willing, again a threefoldness; in science, art and religion we can find a further parallel.

People who are acquainted with anthroposophy sometimes will remark: 'in anthroposophy everything is threefold; ultimately they will even want to describe the

Trinity as threefold!' – My answer is then: 'that is undoubtedly true, but I would prefer to turn things around and say that the Trinity does not present a picture of threefoldness, but that threefoldness is an image of the Trinity.' Thinking this through to its logical end this would mean that the human being was created 'after God's image.' Many of us might find this a worthy conclusion.

Coming back to our subject, the various realms where threefoldness appears can easily be placed side by side. For instance, when we take science, art and religion, it is not so very hard to think of the soul forces of thinking, feeling and willing in that sequence, as also the physical realms of nerve-sense system, rhythmic system and metabolic-limb system.

But now – surprise! In *Towards Social Renewal* and especially also in GA 192[*] Rudolf Steiner speaks about the threefoldness of the social organism as consisting of spiritual-cultural life, the life of rights, and economic life. Generally one would tend to think that the spiritual-cultural life would relate to the nerve-sense system (thinking), the life of rights with the rhythmic system (feeling), and the economic life with the metabolic-limb system (willing). Rudolf Steiner, however, indicates that this economic life must constitute an autonomous member within the social organism, as relatively autonomous as is the nerve-sense system in the human organism. The economy is concerned with all aspects of the production, circulation and consumption of commodities.[12]

When we compare the role of the economic life in today's human society with that of the nerve-sense sys-

---

[*] See bibliography.

tem, and take as our point of departure the conventional view that motor nerves transmit orders from the brain to the muscles, as is usually assumed, we will discover the following striking correspondence.

In the beginning of this study we used the picture that according to the view of modern physiology the body could be considered as a puppet of the nerve-sense system. And what is the role of the economy in modern human society? Is it not a fact that our entire society is more and more running the risk of becoming a puppet of the economy? Here is just one example (from *De Telegraaf* – a leading Dutch newspaper – August 30, 1984):

Germans in Fear of Gas from Particle Board

The chemical formaldehyde, possibly a cause of cancer, has become the focal point in West-Germany of the fiercest discussion of all time regarding economic growth and environmental protection. Things got going accidentally when the City of Wiesbaden ordered a scientific institute to perform measurements in its 22 kindergartens. This happened in reaction to complaints about an abnormal number of instances of bronchitis and other diseases of the air passages. Three kindergartens had to be closed immediately because the air in the classrooms contained seven times the permitted amount of formaldehyde. The gas was shown to come from the particle board with which all the walls and ceilings of the classrooms were lined, and of which the furniture was made. Moreover, the cleaners had worked with a cleaning agent that also contained formaldehyde.

The excitement really rose to a pitch when it became known that last spring already the Minister of Health had a report in his possession containing the warning that formaldehyde possibly causes allergies, brain damage and cancer. The minister was accused of not publishing the report because he wanted to protect the producer of formaldehyde, the chemical giant BASF, from dire consequences for the German economy. Of course, he categorically denied this.

BASF, however, was more than willing to demonstrate the consequences a ban on formaldehyde would have for the German economy. A spokesman for the company mentioned 'DM 300 billion per year'. That is approximately a quarter of the gross national product of West-Germany. According to the spokesman formaldehyde is used in more than 30 different industries and at least three million workers would have to be laid off...

Everywhere, be it in the arts, education, medicine or the rights life, everywhere we run into problems of the economy. Although of course the contents of these disciplines by themselves usually have nothing to do with the economy as such, in the end, the realization of all plans and initiatives is currently dependent on it.

It is evident that in the area of social life we need a totally new way of thinking and feeling.

The title of this study is: *How does the human being move? – The problem of the motor nerves.* The impetus for the study was the statement by Rudolf Steiner that social life cannot be developed until people have corrected the absurd idea of the existence of motor nerves. It is therefore now our task to ask ourselves: where is the connection between the erroneous idea about a particular part of human physiology and the inability to develop an appropriate social life?

We can express the problem we are dealing with in a similar way. According to the physiology of movement we are in fact puppets of our nervous system; as we have seen, that is a bad thought. In practical life we are puppets of the economy. That is a bad fact.

When we rethink the origin of our movements in the way we have described, we will have to turn our thoughts and feelings in a new direction. The experience of thinking in such a way that not only our head is engaged but also our heart can become a source both of joy and of reflection. Only then can we learn to view Rudolf Steiner's remarks in the right light.

Certainly, it is not self-evident that humanity can only build an appropriate social life after it has recognized that the idea of the motor nerves has to be overcome. For this reason it is so important to feel how the train of thought described here can lead us to a totally different orientation relative to the human being as a whole. And here, the all-important thing to realize is that this new orientation is not being given to us – we must develop it out of ourselves.

Consequences

Whoever becomes acquainted for the first time with the ideas developed here will, as was mentioned before, certainly wonder how a change in our picture of the origin of our movements, in other words of the existence of motor nerves that 'direct' our muscles, could possibly have such consequences as are sketched here.

We spoke in the beginning of an East-West polarity, and in this context we involuntarily think of the two giants of today [1980's], Russia and America. We also have to realize how these two are examples of the concepts of 'ideology' and 'economy', and we have characterized the East-West polarity as 'striving for power' and 'striving for wealth' respectively. In the background was the striving for a Middle where the two opposites might be able to meet.

The idea of the motor nerves, as it is generally believed and was described here, is completely based on a materialistic way of thinking. As soon as we have the courage to realize that the life in our muscles is part of a spiritual reality, in which the human being lives also when he is awake, we have made a first step to lift ourselves out of this one-sidedness.

When we introduced the idea that our muscles perceive, we meant at the same time that they have a life of their own. To describe this life is not easy, but that goes for any attempt to describe life in any form whatsoever. We always speak of the life of a plant, the life of an animal, etc. But what do we mean when we say that? The concept of life in a plant is reduced to biochemical processes which demonstrate the phenomena of life as 'function'. This is of course also true for the life of animals, but the explanation of perception by the animal, as it was discussed in this study, is already more difficult to accept for the modern scientist. When we speak of perceiving most of us spontaneously think of something that does the perceiving, in other words, of something we connect with the concept of 'soul life'. Materialistic science has broken with this concept of 'soul'. The word 'perception' is often reduced to irritation, in other words, a reaction to a certain provocation or stimulation.

If we want to extend this thought to the spiritual life of the human being, I have to confess that I experience the word 'irritation' as an extremely meager expression for the activity of soul life. Life, soul and spirit are spiritual realities that reveal themselves in a material world, but are not recognized as such in their own lawfulness. This means that when we speak of muscles that perceive, we point to a soul element which needs to be understood to be just as real as the visible forms of the material life in which we live. To that extent, we would come to correctly value the insight that letting go of the concept of motor nerves immediately leads to the recognition of the reality of soul life in the muscles.

The current East-West polarity arose in the course of centuries, in part because people connected themselves more and more with the world of matter. Therefore, it is my opinion that a total change in the way we view the origin of our movements, as advocated here, is only possible to the extent that we gradually free ourselves of our progressively more deeply rooted materialistic prejudices.

The influence of these thoughts must not be underrated. It has taken humanity centuries to develop the current materialistic picture of the background of our movements. And because this picture was so purely materialistic it went hand-in-hand with the development of conditions in all domains that created the danger that humanity would lose sight of the reality of life, soul and spirit.

No matter how small the realm in which we have looked for the origin of our physical movements, the change in human thinking and feeling that can result from this must not be underrated. Imagine facing a world in which millions of people think about motor nerves in the way this is only possible in a materialistic view of life, in which the awareness of the non-visible is lost more and more. Imagine that facing this world stands the thought that the correction of the picture of motor nerves, as discussed here, is only possible if the concepts of life, soul and spirit are accepted again. This means the birth of a living scientific picture.

It should be remembered that the correction of the hypothesis of motor nerves will influence the entire soul life of humanity. Most of all, however, the point will be to come to an insight into the threefoldness of the human form, the threefoldness of the social organism and, finally, the threefoldness of the structure of the world. Only if we can learn to see the big picture of connections of this kind can we come to recognize the true import of Rudolf Steiner's statement.

Readers who know Rudolf Steiner's Foundation Stone Meditation will be able to appreciate even better the sentence *"Let from the East be enkindled what through the West takes on form*[13] when it is connected with the problem we have dealt with here. Again, we spoke of the East as the realm of power, and of the West as that of wealth, i.e. the economy. Thinking about what would have to happen for a true social relationship to grow between these two powers, we could imagine the substitution of the word 'strength' for 'power'. The consciousness of the West, which is closely dependent on the laws of the economy, would be able to provide social life in the world with an organizing form. Then a space could arise for a Middle between the two polarities.

To just hear what Rudolf Steiner had to say, that the motor nerves are supposed to be sense nerves, and then tell it to others simply does not suffice. It is necessary to completely penetrate his paradoxical statement before a profound change in our comprehension, our consciousness can take place. This is only possible if we are willing to develop a completely new view of our world, a view that lives not only in our head but also in our heart.

In the words of 'The Little Prince': "*It is only with the heart that one can see rightly; what is essential is invisible to the eye.*"[14]

Afterword

By Douglas J. Roger, M.D.

During my training to become an orthopedic surgeon, the concept of a motor nerve was deeply ingrained. The reasons for this relate to surgical complications and the safety of the patient. Most anatomical structures that one might encounter during orthopedic surgery are resilient. For example, if during a surgical procedure you accidently break a bone, it will generally heal if treated properly. If you damage a tendon during surgery, you can usually repair it and it will recover. Injuries to blood vessels are a little more serious, but generally they can be repaired and restored to function. However, the nerves are different, especially the motor nerves. If they are cut or damaged, there is a substantial risk that they might never recover, even if they are surgically repaired. If this occurs, permanent loss of movement (paralysis) can occur.

In this light, Rudolf Steiner's statement that "there are no motor nerves" is astonishing. It is a comment that I have struggled with for many years, since it seemed to me that I had seen motor nerves thousands of times while performing surgical procedures. Without considering the source of such a comment, it might be tempting to dismiss it as fantasy or fiction. However, anyone familiar with Rudolf Steiner's' work cannot dismiss such a comment so easily. He frequently left riddles whose contemplation was part of his cognitive method which espouses a schooling of consciousness, rather than an accumulation of facts. The contemplation of these riddles usually follows a familiar sequence. At first, only strained and awkward explanations can be considered. This is usually followed by a tentative conclusion, which sits uneasily. In many cases, we are left to hold the question, without the benefit of a satisfactory answer, for long periods of time, sometimes years. With patience, persistence, and a modicum of grace, enlightenment may finally come. This is the type of approach that is best suited for the question of the motor nerves.

Centuries ago a materialistic model of the brain and nerves was proposed and accepted by the scientific community. The brain was the "command center." It received information from the "sense nerves" and gave "commands" through the "motor nerves". This places the brain at the pinnacle of the human organism, and results in a one-sided and distorted view of the human being. In contrast, the spiritual scientific view of man encompasses a threefold concept of man. Rather than the brain and nerve sense system as the dominant force of the human organism, it contemplates two other important aspects of the human being, namely, the rhythmic system and the metabolic-limb system that encompasses a muscle being. The concept of a "muscle being" is what concerns us here, and I use this term in the same sense that Dr. Mees uses the term "Muscle Man." This muscle being of man has an independent yet interwoven relationship with the other aspects of the human being, specifically with the nerve sense being of man.

This muscle being lives simultaneously in two different worlds. On the one hand is the visible world, which is intimately connected to the nerve sense being of man. It is from this relationship that I can physically react to my sense impressions and my conscious thinking life. It is also from this relationship that I can be aware of my individual physical movements. The nerve sense being of man is connected to the muscle being of man through the "motor" nerves, but there is also a reciprocal connection that emanates from the muscle being of man and goes to the nerve sense being of man, as we shall see.

The second aspect of the muscle being of man is its connection to karma and destiny. This connection is unconscious to all except to those that posses the faculty of Intuition. This aspect of the muscle being of man is responsible for those movements and activities in life that are not entirely conscious. Those unexplained activities such as why we did not go to the store but to the park, where we met someone who had a profound impact on our lives, or why we went for a walk on the beach instead of a hike in the mountains, and all that results from activities of this kind, are part of the unconscious will activity in our muscle being. These activities are not actively part of our nerve sense being, but are directed from the unconscious activity of the muscle being in us. Rudolf Steiner speaks about the muscle being in us as the entity that is the architect of many accidents and tragedies in our life. These actions can create events and circumstances which help to facilitate the process of soul transformation that is directly related to our individual karma and destiny.

Characterizing the relationship between the nerve sense being of man and muscle being of man is the key to understanding Steiner's enigmatic statements regarding the motor nerves. The dance imagery of Dr. Mees is particularly helpful in this regard, and it also opens the doorway to the social question which relates to the motor nerve problem. Any experienced ballroom dancer will tell you that there must be a leader and there must be a follower. The leader does not tell his partner "this is what you must do." Rather, the leader says to his partner through subtle gestures and signals, "this is what I am doing." The follower then responds, not out of being commanded, but out of a collaborative serving impulse, to what the leader is doing. In addition, there are circumstances in ballroom dancing where the follower will send signals to the leader based on what is happening on the dance floor. This dancing interaction can serve as an image of the ideal relationship not only between the nerve sense being of man and muscle being of man, but also between the economic sphere and the cultural spiritual sphere in the threefold social order.

The traditional picture of a motor nerve involves an "efferent" motor impulse. This motor impulse is a "command" signal from the brain to the muscle. This is the modern, scientific concept of the alpha motor neuron. However, as Dr. Mees has shown, this actually is not a motor impulse to the muscle, but rather a sensory impulse to the muscle being of man. It is the way that the nerve sense being of man informs the muscle being of man about what it is doing. The brain in the nerve sense being of man reflects the activity of the soul of the human being (*Human and Cosmic Thought*, GA 151, lecture 4), and the soul acts as an intermediary between the body and the "I", which is a spiritual, non-physical entity. The muscle being of man responds, much like the follower in ballroom dancing, by mirroring the spiritual activity of the soul with the physical movement of the body. This is why gesture is an outer image of the inner soul activity of man (*Apocalypse of Saint John*, GA 104, lecture 4).

When Dr. Mees wrote about the motor nerves, the more modern concept of proprioception had not yet been extensively developed. Proprioception is the ability to perceive the position and movement of the parts of one's own body, as well as the strength and effort which are required to move the parts of the body. From an anthroposophical perspective, one might say that proprioception is the process by which the nerve sense being of man becomes informed about the activities of the muscle being of man. This perception of the muscle being by the nerve sense being includes the position, movement, and effort that the muscle being actively accomplishes through the muscle activities that move the skeleton. The proprioceptive feedback from the muscle to the spinal cord and brain is not a speculative theory. It is a highly studied area of modern medicine that uses microelectrodes and elaborate, highly sophisticated equipment to study "afferent" impulses from the periphery to the central nervous system.

The proprioceptive feedback mechanism from the muscle being to the nerve sense being of man involves the muscles, tendons, and joints, and there are very specialized microscopic structures that facilitate this perception. For example, there are muscle spindles, which measure direct muscle function, Golgi tendon organs, which measure tension and exertion in the myotendinous region of the muscle tendon unit, and Ruffini endings, which are specialized nerves in the joints and surrounding tissue which monitor and report the position of the joint in space. These signals travel along the neurons of the "afferent" pathway, that is, from the muscle to the spinal cord and brain. However, these afferent neuron pathways (axons) are located within the so-called motor nerves, even though the signals are travelling opposite to the direction of the alpha motor neuron impulse.

Motor nerves therefore do not have a nerve signal that travels only one way from the brain toward the muscle. Rather, they have nerve impulses that travel in both directions, from the brain to the muscle and also from the muscle to the brain. There are, of course, interconnected signals from the muscle to the spinal cord and back to the muscle such as a typical myotactic reflex. A simple phenomenological example of this is the knee jerk reflex. The brain and the spinal cord need not be contemplated separately here, since we are considering them together as part of the nerve sense being of man. In addition, there is very little activity in the spinal cord that is not also transmitted to the brain. Thus, the essence of the relationship between the nerve sense being of man and the muscle being of man can truly be viewed as a two-way dialogue, not a one-way command. When the stretch receptor is activated through the tapping of the stretched patellar tendon, the first impulse is the afferent impulse from the muscle tendon unit to the spinal cord (and also to the brain). The second impulse is the efferent impulse from the spinal cord to the muscle, and this results in the "jerk" of the knee. There is a third impulse (afferent) that goes from the muscle to the spinal cord and the brain, which informs the spinal cord and brain that a movement of the muscle has in fact occurred. The nerve sense being of man thus becomes aware of our movements as a result of this process of proprioception. Nevertheless, we are not aware of how we actually move. We say to ourselves "I am going to take a sip of tea." We do not say to ourselves "I am going to flex my brachialis muscle, relax my biceps a little, activate most of my pronator teres muscle and increase the tone of my interosseous muscles." The mechanism of how we actually move approaches the unconsciousness realm of the muscle being of man. This is the realm of the human will.

The will from this perspective in the human being works in two aspects, one in the nerve sense being and one in the muscle being. The will in the nerve sense being is where we make cognitive and conscious decisions through our thinking. This is also where we will our thinking in the sense described in Rudolf Steiner's *Philosophy of Freedom* (GA 4). We are conscious of what we think and we are free in our higher aspect to choose what we think about. The will in its other aspect is expressed in the muscle being of man. When we move our arm or take a walk, we see the effect of the will working directly on the metabolism in our muscle being, and the nerves perceive this occurrence, through the process of proprioception, as we have seen. We can therefore become conscious of the activities of our will, both in the nerve sense being of man, and in the muscle being of man, and this leads us to the social question.

The riddle of the social question in relation to the motor nerves begins with the modern picture of the human organism as a puppet of the nerve sense system. The correlation of this is the social organism as a "puppet" of the economic sphere. To the extent that the communication of the economic sphere involves a command to the social organism, rather than a dialogue with the other members of the social organism, especially the spiritual cultural life, the social organism cannot be viable and healthy.

There is a similar concordance when the new way of considering the "motor nerves" is considered side by side with the new way of looking at the social organism. The "motor nerves" must not be part of a command structure, whereby the human being is a puppet of the brain. Rather, the human being must be viewed with the muscle being of man clearly in mind. This muscle being serves my I, but is not a slave to it. The dialogue aspect of this is critical because as we have seen, the muscle being has access to realms of karma and destiny that are closed off to the brain and the nerve sense being of man. The brain and nerve sense being of man must learn to "listen" to the muscle being of man and to dialogue with it, rather than to command it.

Similarly, in the social organism the economic sphere must not become a rigid taskmaster to the whole of the social organism. It must not "command" as a "motor nerve" commands. It must instead participate in a dialogue with the spiritual cultural aspect of the social organism. It must "listen" to this aspect, and cultivate a new social dialogue. In this way, the entire social organism can gain access to realms that are accessible to the spiritual cultural sphere, but which are so clearly not perceptible to the economic sphere. This access to higher realms is essential for the social organism to go into the future with a healthy living vitality. It is precisely this access that cannot be achieved without a dialogue structure, rather than a command structure, between the economic sphere and the cultural spiritual sphere of the threefold social order.

Addendum I

Fundamentals of Therapy by Rudolf Steiner, PhD and Ita Wegman, MD, GA 27

CHAPTER II
WHY MAN IS SUBJECT TO ILLNESS

Anyone who reflects on the fact that the human being can be diseased, will find himself involved in a paradox which he cannot avoid if he wishes to think purely on the lines of Natural Science. He will have to assume to begin with that this paradox lies in the very nature of existence. For, outwardly considered, whatever takes place in the morbid process is a process of Nature. But that which replaces it in health is also a process of Nature.

In the first place, the processes of Nature are known to us only by observation of the world external to Man, and of Man himself inasmuch as we set to work observing him in just the same way as we observe external Nature. In doing so, we conceive him as a piece of Nature. We conceive that the processes going on within him, however complicated, are of the same kind as the processes we can observe outside him—the outer processes of Nature.

Here, however, a question emerges which is quite unanswerable from this point of view. How do there arise in Man (not to speak, at this point, of the animal) processes of Nature which run counter to the healthy ones?

The healthy human body would seem to be intelligible as a piece of Nature; not so the diseased. It must, therefore, in some way be intelligible out of itself, by virtue of something which it does not have from Nature.

The prevalent idea is that the conscious mental life in Man has for its physical foundation a very complicated process of Nature—a further elaboration of the processes we find outside him. Let us however look and see: *does* the continuation of any process of Nature, upon the ground of the healthy human body, ever call forth conscious experience as such? The very reverse is the case. Consciousness is extinguished when Nature's process is continued in a straightforward line. This is what happens in sleep; it happens, too, in faintness.

Consider on the other hand how the conscious mental life is sharpened when an organ becomes diseased. Pain ensues, or—at the least—discomfort and displeasure. The life of sentient feeling receives a content which lacks in ordinary life; the life of Will on the other hand is impaired. The movement of a limb which takes place as a matter of course in the healthy state can no longer be accomplished properly; pain or discomfort hinders and prevents it.

Observe now the transition from the painful movement of a limb to its paralysis. In the movement accompanied by pain we have the initial stages of a movement paralysis. The conscious activity of the human Spirit takes hold of the body. In health this activity reveals itself to begin with in the life of thought or ideation. We actuate a certain idea, and the movement of a limb ensues. We do not enter consciously with the idea into the organic processes which culminate in the movement. The idea dives down into the unconscious. Between the idea and the movement an act of feeling intervenes; but this—in the healthy condition—works in the soul only, it does not attach itself distinctly to any bodily organic process. In disease however, it is different. The feeling, experienced in health as a thing distinct and apart, unites with the physical organization in the conscious experience of illness.

The healthy processes of feeling and the conscious experience of illness thus appear in their relationship. There must be something there, which, when the body is in health, is less intensely united with it than when it is diseased. To spiritual perception this "something" is revealed to be the astral body. The astral body is a supersensible organization within that which the senses can perceive. If it takes hold of an organ but loosely, it leads to an inner experience of soul—an experience which subsists in itself and is not felt to be in connection with the body. If, on the other hand, the astral body takes hold of an organ strongly or intensely, it leads to the consciousness of illness. One of the forms of illness must indeed be conceived as an abnormal seizure of the organism by the astral body. This form of illness causes the spiritual part of man to dive down into the body more deeply than is the case in health.

Now thinking too has its physical foundation in the body. In health however, it is still more loosely connected, still freer of the bodily foundation, than the life of feeling. Spiritual perception finds beside the astral body a special Ego-organization which lives and expresses itself with freedom of soul in thought. If with this Ego-organization man takes intense hold of his bodily nature, the ensuing condition makes his observation of his own organism similar to that of the external world. When he observes a thing or process of the outer world, the thought in man and the object observed are not in living mutual interplay; they are, in a sense, independent of each other. In a human limb this condition only takes place when it is paralyzed. The limb then becomes a piece of the outer world. The Ego-organization is no longer loosely united with it as it is in health, when it can unite with the limb in the act of movement and withdraw again at once. It dives down into the limb permanently and is no longer able to withdraw.

Here again the processes of the healthy movement of a limb and of paralysis stand side by side in their relationship. **Nay more, we recognize distinctly: The healthy act of movement is a paralysis in its initial stages—a paralysis which is arrested as soon as it begins.**

We must see the very essence of illness in this intensive union of the astral body or Ego-organization with the physical organism. Yet is this union only an intensification of that which exists more loosely in a state of health. Even the normal way in which the astral and Ego-organization take hold of the human body, is related not to the healthy processes of life, but to the diseased. Wherever the soul and Spirit are at work, they annul the ordinary functioning of the body, transforming it into its opposite. In so doing they bring the body into a line of action where illness tends to set in. In normal life this is regulated directly it arises by a process of self-healing.

A certain form of illness occurs when the Spirit, or the soul, pushes its way too far into the organism, with the result that the self-healing process can either not take place at all or is too slow.

In the faculties of soul and Spirit, therefore, we have to seek the causes of disease. Healing must then consist in loosening this element of soul or Spirit from the physical organization. This is the one kind of disease. There is another. The Ego-organization and the astral body may be prevented from reaching even that looser union with the bodily nature which is conditioned, in ordinary life, by the independent activities of Feeling, Thought and Will. Then, in the organs or processes which the soul and Spirit are thus unable to approach, there will be a continuation of the healthy processes beyond the due measure which is right for the organism as a whole. But spiritual perception shows that in such a case the physical organism does not merely carry out the lifeless processes of external Nature. For the physical organism is permeated by an etheric. The physical organism alone could never call forth a process of self-healing; it is in the etheric organism that this process is kindled. We are thus led to recognize health as that condition which has its origin in the etheric. Healing must therefore consist in a treatment of the etheric organism.

Addendum II

PASSAGES BY RUDOLF STEINER ON MOTOR NERVES NOT PUBLISHED IN ENGLISH
Translation from German by William Lindeman

Geisteswissenschaftliche Behandlung sozialer und pädagogischer Fragen, GA 192

GA 192, pages 51-53, lecture by Rudolf Steiner, April 23, 1919

Thus you can see that, as human beings, we are in a peculiar situation today. Natural science has taken many steps forward, and has influenced our thought habits in such a way that basically all the social thinking of those who think socially is natural-scientifically oriented, even if they don't know it. This natural science is not capable of judging the human being in the right way. For example, it propounds the blatant nonsense that when you feel something, the feeling is mediated by the nervous system. That is pure nonsense. The feeling is directly mediated by our breathing system, by the rhythmic system, just as a thought is mediated by the nerve-sense system. And the will is mediated by the metabolic system, not at all by the nervous system as the basic element. Only the thought of the will is mediated by the nervous system.

Only through the fact that, as human beings, you have a clear consciousness of your will activity is the nervous system involved. By your thinking along with your will activity, the nervous system is involved. Because one does not know this, that awfully misleading view has arisen in today's physiology and anatomy that one distinguishes between sensory nerves and motor nerves.

There is a no more blatant incorrectness than this distinction between sensory nerves and motor nerves in the human body. The anatomists are always in difficulty when they discuss this matter, but they never get out of it. They are in such awful difficulty because anatomically these two kinds of nerves are not different from each other. To claim that they are is pure speculation. And everything connected with the study of tabes [a gradually progressive emaciation] absolutely does not hold up. Motor nerves are not different from sensory nerves, because motor nerves are not there in order to set our muscles into motion. Our muscles are set into motion through our metabolism. And whereas you perceive the outer world with the so-called sensory nerves in a roundabout way through the senses, you perceive with other nerves your own movements, the movements of your muscles. Only erroneously does today's physiology call them motor nerves.

Awful assumptions like this exist in science. They corrupt what passes over into popular consciousness and works in a much more corrupting way than one usually thinks.

So, natural science is not far enough along to gain insight into this threefold human being. In natural science one can wait to see whether theoretical views become popular a few years earlier or later. This doesn't affect people's happiness. But the thinking needed to comprehend this threefold human being is not present. That way of thinking must be present, however, in order to understand the social organism in its threefoldness. Then things become serious. We stand today at the point in time where these things *must* be understood. Therefore, this kind of a reversal of thinking, this kind of a relearning, is truly necessary, not only for the naïve person, but most of all for learned people. Naïve people at least know nothing of all that has been set up in natural science to hide, unconsciously, the threefoldness of man. Learned people, however, are stuffed with all the concepts that today make this threefoldness out to be nonsense. For the physiologist of today, threefoldness is total rubbish. He will strongly object if one says to him that there are no motor nerves, if one speaks about the fact that feelings are not just as much mediated by the nervous system as are thoughts, but rather that only the *thought* of the feeling is mediated by the nerve, the *consciousness* of the feeling, not the feeling as such. One knows well the objections to these things. Of course people can say: "All right. Look. You perceive musical things. You perceive them through the senses." No! Our feeling for music is present in a much more complicated way. It is based on the fact that the breathing rhythm in the brain meets the sense perception, and in the striking together of the breathing rhythm with the outer sense perception, the musical-aesthetic feeling arises. Also there it is the case that the basic element lies in the rhythmic system. And the nervous system is what brings the basic element to consciousness.

GA 192, pages 153-155, lecture by Rudolf Steiner, June 8, 1919

Another horrible mental picture lives in our official science (i.e., in a science believed in everywhere on authority). This science partakes of the idolatrous worship of everything that has arisen in modern times as a so-lofty culture. When it wants to express something with special mysteriousness, why shouldn't this modern science take recourse to what it happens to worship most? So now the nervous system has become for it a sum of telegraph lines; our entire human nerve activity has become for it a remarkably complicated telegraphic functioning. The eye perceives. The skin perceives along with it. What is perceived from outside is conducted to the telegraph-station brain by sensory nerves. Then, sitting there in the brain, is I don't know what kind of an entity (modern science of course denies any kind of spiritual being). Through this entity, which has become a catch phrase because one sees nothing real in it, the perceptions of "sensory" nerves are changed into will movements by "motor" nerves. And the difference between sensory and motor nerves is pounded into the young person; and one's whole view of man is built upon this difference.

For years now I have fought against this absurdity [*Unding*] of the separation between sensory and motor nerves. First of all because this distinction is an absurdity, for, the so-called motor nerves exist to do nothing other than what sensory nerves also do. A sensory nerve, a sense nerve, exists as an instrument for us to perceive what takes place in our sense organization. And a so-called motor nerve is not a motor nerve; rather, it is a sensitive nerve; it is there only so that I can perceive my own hand movements, my own movements, which come from sources other than motor nerves. Motor nerves are inner sense nerves for perceiving the initiatives of my own will. In order for me to perceive the outer world, to perceive what occurs in my sense apparatus, sensory nerves exist. And so that I do not remain an unknown entity to myself when I walk, strike, or grasp without knowing anything about it, the so-called motor nerves exist. So motor nerves are not there to activate our will, but rather so that we can perceive what the will does in us. Everything that has been formulated out of the misguided, intellectual knowledge of our time is a real scientific absurdity. That is one reason why I have fought this absurdity for years now.

But there is another reason why this absurdity must be rooted out, this superstition about sensory and motor nerves, between which there is no other difference than that one is sensitive to what is outside, and the other to what is in one's own body. This other reason is as follows.

No one, in any social science, can gain a right understanding of man's relation to work who builds his concepts, his mental pictures, upon the misguided distinction between sensory and motor nerves. For, one will always gain quite curious concepts about what human work actually is when one asks on the one hand: "What is actually transpiring in a human being when he works, when he brings his muscles into movement?" and on the other hand has no inkling of the fact that this bringing-his-muscles-into-movement is not based upon the so-called motor nerves, but rather upon the direct involvement of the soul with the outer world. I can of course only point out these questions to you, because still today not even the most primitive mental pictures are present for grasping the point. People still understand absolutely nothing about these things, because our education has still not yet brought into motion even the most primitive mental picture for understanding such things, because it continues to work with the insanity of distinguishing between sensory and motor nerves.

When I come in touch with a machine, I must come in touch with it as a whole human being. Above all, I must establish there a relation between my muscles and this machine. This relation is what human work really rests upon. This relation is the crucial thing when one wants to evaluate work in a social way. The crucial point is the completely special relation of man to the basis of work.

What is then the concept of work that we work with today? [It is the following.] What transpires in man when, as we say, his work does not differ when he labors at a machine, or chops wood, or engages in sports for pleasure. He might just as well wear himself out with the pleasure of sports. He can consume just as much work strength in socially superfluous sports as in socially useful wood chopping. And it is the illusion about the difference between motor and sensory nerves that psychologically diverts people from grasping a real concept of work, a concept that can be grasped only when one does not look at how he wears himself out, but rather at the way he places himself in a relation to his social surroundings.

I believe you, that you have not yet received a clear concept about this, because the concepts that one can receive today about these things are so twisted by our education that it will take some time before one will find the transition from the socially nonsensical concept of work, from the insane scientific concept of the distinction between sensitive and motor nerves. But at the same time in such things lies the reason why we think so impractically. For, how can a humanity think practically about what is practical when it gives itself over to the following insane mental picture: In our inner life, a telegraphic apparatus holds sway, and the wires go to something or other in the brain and are switched there over into other wires, into sensitive and motor nerves. Our inability to think in a really social way originates from our non-science, which springs from a twisted education, and is believed in by the broad public that is led astray by the plague of newspapers.

Addendum III

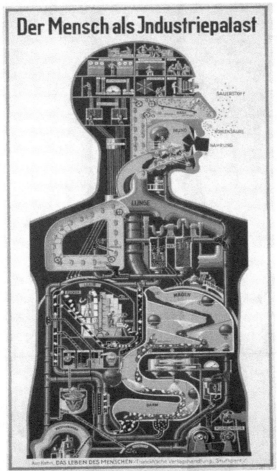

To view an animated version of this piece of art by Henning M. Lederer go to the web site www.vimeo.com/6505158.

Notes

1 Rudolf Steiner, *Towards Social Renewal*, GA 23
2 Rudolf Steiner, *Geisteswissenschaftliche Behandlung sozialer und pädagogischer Fragen*, GA 192, See Addendum II below
3 Ibid
4 Ibid
5 From Wikipedia, the free encyclopedia: Anthroposophy, a philosophy founded by Rudolf Steiner, postulates the existence of an objective, intellectually comprehensible spiritual world accessible to direct experience through inner development. More specifically, it aims to develop faculties of perceptive imagination, inspiration and intuition through cultivating a form of thinking independent of sensory experience, and to present the results thus derived in a manner subject to rational verification. In its investigations of the spiritual world, anthroposophy aims to attain the precision and clarity attained by the natural sciences in their investigations of the physical world. Anthroposophical ideas have been applied practically in many areas including Steiner/Waldorf education, special education (most prominently through the Camphill Movement), agriculture, medicine, ethical banking, organizational development and the arts.
6 Rudolf Steiner, *Geisteswissenschaftliche Behandlung sozialer und pädagogischer Fragen*, GA 192
7 Ibid
8 "Der Mensch als Industriepalast" by Dr. Fritz Kahn, see Addendum III
9 Pierre Teilhard de Chardin *The Future of Man* Chapter 10

The actual quote: "The animal knows, it has been said; but only man, among animals, knows that he knows."

10 Rudolf Steiner, PhD and Ita Wegman, MD *Fundamentals of Therapy*, GA 27. Quote found in Chapter II "Why Man is Subject to Illness" See Addendum 1 below.

11 Rudolf Steiner, *Geisteswissenschaftliche Behandlung sozialer und pädagogischer Fragen*, GA 192, 7th Lecture June 8, 1919

12 Rudolf Steiner, *Towards Social Renewal*, GA 23

13 Rudolf Steiner, *The Foundation Stone Meditation*, Anthroposophical Society in America 1998

14 Antoine de Saint-Exupéry, *The Little Prince*

Bibliography

Documentation of the passages in Rudolf Steiner's work in which he spoke about the problem of the 'motor nerves':

Background to the Gospel of St. Mark, GA 124, March 7, 1911

Human and Cosmic Thought, GA 151, January 23, 1914

Geisteswissenschaftliche Behandlung sozialer und pädagogischer Fragen, GA 192, April 23, June 8 & June 9, 1919 (not in English)

Karma of Vocation, GA 172, November 6, 1916

Riddles of the Soul, GA 21, 1917, Chapter IV/6

May 21, 1917 München unveröffentlicht: Der motori-sche Nerv nimmt den unmittlelbar bewirkten Stoffwechsel wahr. Allen Bewegungsprozessen liegen Stoffwechselprozesse zugrunde.

Introducing Anthroposophical Medicine, GA 312, March 23 & April 9, 1920

Renewal of Education, GA 301, April 21, 1920

Curative Eurythmy, GA 315, April 14, 1921

Geisteswissenschaftliche Gesichtspunkte zur Therapie, GA 313, April 17, 1921 (published in English as *Anthroposophical Spiritual Science and Medical Therapy* by Mercury Press, Chestnut Ridge, NY)

Other Sources:

Rudolf Steiner, *Towards Social Renewal*, GA 23

Rudolf Steiner & Ita Wegman, *Fundamentals of Therapy*, GA 27

Rudolf Steiner, *Between Death and Rebirth*, GA 141

Antoine de Saint-Exupéry, *The Little Prince*